Walnara Arnaud Moura Formiga
Antônio Vitor Machado
Patrício B. Maracajá

Incorporating Bee Honey into School Meals in São Domingos-PB

Walnara Arnaud Moura Formiga
Antônio Vitor Machado
Patrício B. Maracajá

Incorporating Bee Honey into School Meals in São Domingos-PB

Feasibility of Incorporating Bee Honey into School Meals at a School in São Domingos-PB

ScienciaScripts

Imprint

Cover image: www.ingimage.com

This book is a translation from the original published under ISBN 978-613-9-61746-3.

Publisher:
Sciencia Scripts
is a trademark of
Dodo Books Indian Ocean Ltd. and OmniScriptum S.R.L publishing group

120 High Road, East Finchley, London, N2 9ED, United Kingdom
Str. Armeneasca 28/1, office 1, Chisinau MD-2012, Republic of Moldova, Europe
Printed at: see last page
ISBN: 978-620-7-69999-5

SUMMARY

Thanks

To God, for His unconditional love, for His infinite goodness, mercy and providence in my life, for giving me the grace to have done a Master's degree;

To all my family for their help and encouragement;

To the Federal University of Campina Grande and the Center for Agrifood Sciences and Technologies for the opportunity to complete this dissertation.

To Professor Dr. Patricio Borges Maracaja for his affection, guidance, encouragement and, above all, for believing in my potential and understanding my difficulties, giving me the chance to achieve a great desire;

To Professor Dr. Antonio Vitor Machado for his attention, availability and suggestions during the execution of this work;

To the Professors of the Postgraduate Course in Agro-Industrial Systems for the precepts they imparted;

To my friends on the Agro-Industrial Systems Postgraduate course (especially Lucicleide, José da Silva Sousa (Zezinho), Vitòria Celestino and Cristiane Reis for supporting me and not letting me give up);

To Fabiano, an agronomist, for his help with the statistical program;

To the Mayor, Odaisa de Cassia, and the Secretary of Education of Sao Domingos-PB, Simone de Cassia, for authorizing and supporting this work;

To the Principal of the Maria Marques de Assis Elementary School, Tânia Maria Morais, for her understanding and indispensable help in carrying out this work;

To all my friends (especially Jakeline Caetano and Maria Fabricia) for always guiding me to choose wisdom.

To everyone who contributed in any way, directly or indirectly, to the completion of this work.

SUMMARY

0his study aimed to promote the introduction of bee honey into school meals at an urban elementary school in Sao Domingos-Pb. The research was descriptive in nature, with a quantitative, qualitative and participatory approach. The study was carried out in a school in the urban area. The population included the 180 students enrolled, the principal, teachers and cooks at the school under analysis, as well as local rural family farmers who sell and/or have sold food for school meals, the mayor and the municipal education secretary. The students were sampled according to Barbetta (2001), so the number of students was equal to 60 children aged between 9 and 13. Two groups were formed for the study, each with 30 students: group 1 was made up of students from the morning shift who received nutritional education about honey, and group 2 was made up of students from the afternoon shift who did not have any educational activity related to the study, questionnaires were administered to the students before and after they were given bee honey in their meals, educational activities were carried out on the benefits of honey with group 1 students and on beekeeping with the farmers, and questionnaires were administered to the teachers, principal, cooks, farmers, mayor and education secretary. The information was analyzed using the statistical software IBM SPSS Statistics, version 21 of 2012 and using Bardin's canteen analysis technique. It was concluded that the introduction of honey is feasible as long as there is nutritional education, since group 1 of students had an acceptability index above the minimum set by the National Education Development Fund (FNDE), while group 2 of students had an acceptability index below the minimum required. The local farmers have shown an interest in producing honey, and the municipal managers have developed real actions that could lead to the permanent inclusion of honey on the school menu and local sustainable development, creating the Municipal Inspection System and establishing partnerships with SEBRAE (Brazilian support service for micro and small enterprises) to encourage the local beekeeping agribusiness.

Keywords: bee honey, school meals, family farming.

1. INTRODUCTION

The school years are a time of great development for students. During this period, many physical and psychological changes occur in their lives, and it is extremely important to have a healthy diet, including at school, as this can favor the learning process.

In order to guarantee learning and improve the quality of life of schoolchildren in Brazil, there is a program known as School Meals. This is thc National School Feeding Program (Programa Nacional de Alimentaçao Escolar - PNAE), which is based on the transfer of financial investments from the Federal Government, on an improvement basis, to the states, the Federal District and municipalities, for the purchase of food for school feeding. Its aim is to satisfy the nutritional requirements of students while they are in the classroom, favoring development, growth, learning, school performance and the formation of healthy eating habits.

In June 2009, with the approval of Law No.^{0} 11.947, it was decided that at least 30% of the amount passed on by the National School Feeding Fund (FNDE) to cover the cost of school meals in Brazilian municipalities and states should be used to buy food directly from rural family farmers and entrepreneurs or their organizations (cooperatives, associations).

Increasing the quality of school meals, developing family farming and stimulating the local economy are all factors that are boosted through family farming in school meals. However, the municipality of Sao Domingos-Pb, like others in the state of Paraiba, still finds it very difficult to acquire products from family farming.

Bee honey is a natural food that consists of a supersaturated solution of glucose and fructose, as well as being rich in micronutrients that are essential for the human body. Because of its composition, this food provides individuals who consume it with antibacterial, anti-inflammatory, energizing and antioxidant properties, among others.

Given the potential benefits that this food can provide to schoolchildren and the fact that it can be produced and marketed locally, as well as the fact that bee honey has not yet been offered in school meals in Sao Domingos-Pb, this work aims to promote the introduction of this food in the local PNAE as a way of promoting social, economic and health benefits for the population of Sao Domingos.

When school meals contain nutritious and healthy foods, they become indispensable to the student's learning and development process, while also guaranteeing a minimum supply of food for needy students. In this context, the introduction of bee honey into the Sao Domingos school diet will enable it to better achieve its principles.

The general aim of this study is to promote the introduction of bee honey as a food in school meals at an urban elementary school in the municipality of Sa

The specific objectives of this project are to carry out nutritional education activities that encourage the consumption of bee honey by schoolchildren; to offer bee honey in sachets as a dessert in a school in the urban area of the municipality in question; Evaluate the acceptance of bee honey by schoolchildren in the urban area of Sao Domingos-PB; Improve the health, learning and development of schoolchildren; Encourage local family farmers to become honey producers; Provide support for advances in the local economy.

2. LITERATURE review

2.1 The importance of food at school.

Food plays a key role throughout an individual's life cycle. Among the different stages of life we can highlight, for example, school age (from 6 to 13 years), which is characterized by a period in which children have a much more intense metabolism when compared to adults, requiring adequate nutrition for their development and learning (PHILIPPI, 2000).

It is at school age that permanent dentition begins, a phase in which good eating and hygiene habits need to be reinforced in order to prevent the occurrence of cavities and/or other infections (ACCIOLY et. al, 2012).

School meals need to contain natural, good quality food, as children at this stage of their lives become more independent, deciding for themselves their tastes, preferences and dislikes, showing a critical sense and these factors will be clearly reflected in the child's general and eating habits (ACCIOLY et. al, 2012).

The lack of food or the use of unhealthy food while students are at school can lead to a decrease in their ability to concentrate and a low response to stimuli, highlighting that the development of healthy eating habits in schoolchildren means that they miss less school and therefore learn more (VITOLO, 2008).

2.2 The National School Feeding Program - PNAE.

2.2.1 Historical Facts

Like any social policy, the PNAE originated from the need to meet a social demand permeated by conflicts and contradictions of interests, so this program emerges as a means of realizing citizenship rights, as described below in part of an official document from the National Education Development Fund (FNDE, 2013), depicting the chronological explanation of the facts related to the program.

The National School Feeding Program, always supported by the grassroots movement, has its origins in the early 1940s, when the then Institute of Nutrition defended the proposal that the federal government should offer school meals. However, it was not possible to implement it due to the lack of financial resources (FNDE, 2013).

In the 1950s, with the drafting of Conjuntura Alimentar e o Problema da Nutriçao no Brasil, a national school lunch program was set up, under public responsibility. From this original plan, only the School Feeding Program survived, with funding from the International Children's Relief Fund (Fisi), now Unicef. In 1955, the School Meals Campaign (CME) was set up, under the Ministry of Education (FNDE, 2013).

From 1976 onwards, although funded by the Ministry of Education and managed by the National School Feeding Campaign, the program was part of the Second National Food and Nutrition Program (Pronan). It was only in 1979 that it was renamed the National School Feeding Program. With the promulgation of the Federal Constitution in 1988, the right to school meals was guaranteed to all elementary school students through a supplementary school feeding program to be offered by the federal, state and municipal governments (FNDE, 2013).

From its creation until 1993, the program was executed in a centralized manner, i.e. the managing body planned the menus, acquired the foodstuffs through a bidding process, hired specialized laboratories to carry out quality control and was also responsible for distributing the food throughout the country (FNDE, 2013).

In 1994, the decentralization of resources to implement the program was instituted through the signing of agreements with the municipalities and with the involvement of the state and Federal District education secretariats, which were delegated responsibility for serving the students in their networks and the municipal networks of the prefectures that had not adhered to decentralization (FNDE, 2013).

The consolidation of decentralization, already under the management of the FNDE, took place with Provisional Measure No. 1.784, of 14/12/98, in which, in addition to the direct transfer to all municipalities and education departments, the transfer was made automatically, without the need to enter into agreements or any other similar instruments, allowing the process to be more agile (FNDE, 2013).

Provisional Measure No. 2.178, of June 28, 2001, brought major advances to the PNAE. These include the requirement that 70% of the funds transferred by the federal government be used exclusively for basic products and respect for regional eating habits and the agricultural vocation of the municipality, encouraging the development of the local economy (FNDE, 2013).

In 2009, Law No.⁰ 11.947 of June 16 and Resolution No.⁰ 38 of July 16, 2009 brought new advances to the PNAE, such as the extension of the program to the entire public basic education and youth and adult education network, and the guarantee that 30% of FNDE transfers (around 900 million reais) be invested in the purchase of family farming products, stimulating the economic development of communities (FNDE, 2013).

Another major achievement was the establishment, in each Brazilian municipality, of the School Feeding Council (CAE) as a deliberative, supervisory and advisory body for the execution of the Program. This happened with another reissue of Provisional Measure 1.784/98, on June 2, 2000, under the number 1979-19. Currently, the CAEs are made up of representatives of organized civil entities, education workers, students, parents and representatives of the executive branch (FNDE, 2013).

From 2006 onwards, a fundamental achievement was the requirement for the presence of a nutritionist as the Technical Responsible for the Program, as well as a technical staff made up of these professionals in all the Executing Entities, which led to a significant improvement in the quality of the Pnae in terms of achieving its objective (FNDE, 2003).

Another milestone worth highlighting, from 2006 onwards, was the establishment of a partnership between the FNDE and Federal Higher Education Institutions, culminating in the creation of the Collaborating Centers for School Food and Nutrition - Cecanes, which are reference and support units set up to develop actions and projects of interest and need to the PNAE, with a structure and team to carry out extension, research and teaching activities (FNDE, 2003).

The PNAE is known worldwide as a successful case of a Sustainable School Feeding Program. In this context, it is important to highlight the International Agreements signed with the United Nations Food and Agriculture Organization - FAO and the World Food Program - WFP, through the Brazilian Cooperation Agency of the Ministry of Foreign Affairs, with a view to supporting the development of Sustainable School Feeding Programs in Latin American, Caribbean, African and Asian countries, under the principles of Food and Nutritional Security and the Human Right to Adequate Food (FNDE, 2003).

In 2009, the passing of Law No.^{0} 11.947, of June 16, brought new advances to the PNAE, such as the extension of the program to the entire public basic education network, including students participating in the More Education Program, and to young people and adults, and the guarantee that at least 30% of FNDE transfers will be invested in the purchase of family farming products. Another important change was the inclusion, in 2013, of assistance for students attending Specialized Educational Assistance (AEE), for those in semi-presential Youth and Adult Education and for those enrolled in full-time schools (FNDE, 2003).

In terms of financial resources, the PNAE **transfers** differentiated funds to meet ethnic diversity and nutritional needs by age group and social vulnerability. In this way, it is worth highlighting the fact that the Program prioritizes agrarian reform settlements, traditional indigenous communities and quilombola communities in terms of the purchase of food from Family Farming, as well as differentiating the *per capita* amount transferred to students enrolled in schools located in indigenous areas and quilombola remnants. In 2012, the amount passed on to students enrolled in nurseries and pre-schools was increased, in line with the government's policy of prioritizing early childhood education (FNDE, 2003).

On June 17, 2013, FNDE Resolution No. 26 was published, which strengthens one of the axes of the Program, Food and Nutrition Education (FNE), by dedicating a Section to FNE actions. This measure is in line with current public policies related to Food and Nutrition Security (FNS), given the existence of the FNS Plan, the National Plan to Combat Obesity and the Strategic Action Plan to Combat Chronic Non-Communicable Diseases (CNCD) (FNDE, 2003).

Finally, it is worth noting that in 2000, the PNAE served around 37.1 million students with an investment of R$ 901.7 million. In 2013, approximately 43 million students were served with an investment of around R$3.5 billion (FNDE, 2003).

2.2.2 FOOD AND NUTRITION IN THE PNAE.

According to FNDE (2013), the following information can be highlighted about food and nutrition actions in the PNAE:

These actions include various activities: assessing the nutritional status of the students served by the PNAE; identifying individuals with specific nutritional needs; carrying out food and nutrition education actions for the school community, in conjunction with the school's pedagogical coordination; planning and coordinating the application of the acceptability test; drawing up and implementing the Good Practices Manual in accordance with the reality of each school unit; interaction with family farmers and rural family entrepreneurs in order to get to know local production, including these products in school meals; planning and monitoring school food menus, among others, with the nutritionist being the technical responsible (RT) for planning, coordinating, directing, supervising and evaluating all food and nutrition actions within the scope of school meals (FNDE, 2003).

The school lunch menu is an instrument that aims to ensure the provision of a healthy and adequate diet, which guarantees that students' nutritional needs are met during the school term and acts as a pedagogical element, characterizing an important food and nutrition education action (FNDE, 2003).

The menus must be prepared by the RT nutritionist, taking into account: the use of healthy and adequate food, including the use of varied, safe food that respects culture, traditions and healthy eating habits, meeting the nutritional needs of students in accordance with their age group and state of health; foodstuffs produced locally, preferably by family farming and rural family entrepreneurs; the time at which food is served and the appropriate food for each type of meal, the macro and micronutrient requirements; the cultural specificities of indigenous and/or quilombola communities; the provision of at least 3 portions of fruit and vegetables per week (200g/pupil/week), with fruit-based drinks not replacing the mandatory provision of fresh fruit; sensory aspects, such as colors, flavors, texture, food combinations and preparation techniques (FNDE, 2003).

2.2.3 THE PNAE AND THE CHANGING NUTRITIONAL PROFILE OF BRAZILIANS IN THE SPHERE OF FOOD PRODUCTION .

In Brazil, over the last few decades, there has been a phenomenon called the "nutritional transition", characterized by an increase in overweight and chronic non-communicable disease rates, combined with a large but declining prevalence of nutritional deficits (KAC; VELASQUEZ; MELÉNDEZ, 2003). In addition, in the sphere of production, a production model is being strengthened that disseminates practices and makes available types of food that are closely linked to this health situation and to environmental degradation. This production is based on intensive, mechanized agriculture, with a high use of chemical products, with environmental and social consequences, such as the marginalization of a large part of rural producers and an increase in poverty in the countryside (FRIEDMANN, 2000).

In order to combat these trends, the state has begun to act in this sector on the basis of structuring models based on the concepts of Sustainable Food and Nutrition Security (SAN). The policies that stem from this approach are based on two basic components: the food component, related to production, availability, marketing and access to food, and the nutritional component, related to eating practices and the biological use of food, and therefore to the state of nutrition of the population (CONSEA, 2004).

Both components propose a more sustainable model of food production and consumption that brings the production of small family farmers and food consumption closer together, helping to reconnect the food chain through a closer relationship between the countryside and the city.

From this perspective, public food programs such as the National School Feeding Program (PNAE) appear as potential reintegrators of these components, given their ability to help tackle the problems of food consumption and production. This is due, on the one hand, to the integration of policies related to the health of the school-age population and, on the other, to the creation of markets for family farmers, including the potential to encourage environmental management practices, with bee honey being an example of a natural and nutritious food capable of giving existence to marketing in particular through appropriate beekeeping activity, encouraging healthy habits of environmental management.

2.3 Family Farming

Family farming is currently considered to be a strategic sector, both for maintaining and recovering employment, distributing income, guaranteeing the country's food sovereignty and building sustainable development (SCHUCH, 2013).

According to Perondi and Ribeiro (2000), family farming is very dynamic, with the ability to combine agricultural and non-agricultural activities, somehow seeking an income outside the productive establishment, in a commercial or service activity.

The importance of family farming in the process of rural development is certain, and its role in occupation and income in rural areas is currently being discussed, as well as its responsibility for the sustainable use of natural resources (FLORES, 2002).

Family farming is now recognized as a social actor. Once seen only as the rural poor, low-income producers or small producers, family farmers are now perceived as the bearers of another concept of agriculture, different and alternative to the landowning and patronage agriculture that dominates the country. The Family Farming Support Program (PRONAF), implemented in Brazil in the 1990s, is an expression of this change (WANDERLEY, 2000).

2.3.1 FAMILY FARMING AND THE PNAE

Law[0] 11.947 requires that of the total financial resources passed on by the FNDE to the executing entities, within the scope of the PNAE, at least 30% (thirty percent) must be used to purchase foodstuffs directly from family farming and rural family entrepreneurs or their organizations, giving priority to agrarian reform settlements, traditional indigenous communities and quilombola communities (FNDE, 2013).

The FNDE has published Resolution 38 (July 16, 2009), which regulates the school feeding law and which, in relation to family farming, states that foodstuffs can be purchased without a call for tenders, as long as the prices are in line with the local market. In this case, the executing agency (education department or school) must publish a public call for tenders in a newspaper or on a widely circulated bulletin board. This call for tenders will define the foodstuffs and the quantity to be purchased based on the menu drawn up by the nutritionist, who is the technical manager (FNDE, 2013).

Those interested in supplying school meals will then have to submit a sales plan (there is a separate form) along with the other documents described in the resolution (FNDE, 2013).

The following are eligible to supply school meals: family farmers and rural family entrepreneurs who have a DAP (Declaration of Aptitude to PRONAF - National Program for Strengthening Family Farming) organized in formal or informal groups. Priority will be given to proposals from municipal groups. This can be supplemented, if necessary, with proposals from groups in the region, rural territory, state or country. Priority should also be given, whenever possible, to organic or agroecological food (FNDE, 2013).

2.4 Bee honey

Bee honey is defined as a food product produced by honey bees from the nectar of flowers or secretions from living parts of plants or from the excretions of plant-sucking insects that remain on living parts of plants, which the bees collect, transform, combine with their own specifics, store and leave to mature in the combs of the hive (BRASIL, 2000).

In Brazil, the Ministry of Agriculture, Livestock and Supply's Normative Instruction No.^{0} 11, dated 20/10/2000, describes in detail the rules for the production, classification, processing, packaging, distribution, identification and quality certification of honey. This regulation defines the composition of honey as a concentrated solution of sugars, predominantly glucose and fructose. It also contains a complex mixture of other carbohydrates, enzymes, amino acids, organic acids, minerals, aromatic substances, pigments and pollen grains which may contain beeswax from the extraction process (BRASIL, 2000).

A major obstacle to increasing consumption is the fact that Brazilians generally consider honey to be nothing more than a useful natural medicine for the respiratory tract. However, it is a food rich in nutrients, as it contains large quantities of sugars and smaller quantities of minerals, organic acids, proteins and vitamins (EBELING, 2002).

According to the Ministry of Health's Food Guide for the Brazilian Population (BRASIL, 2005), honey belongs to the group of sugars and sweets. The diet of school-age children should contain a daily portion of these foods, which should provide around 5% of the daily calories, which represents, for eutrophic children, around 110 calories/day and to provide 110 calories, 37.5g of honey is needed, which is equivalent to 2 ½ tablespoons.

2.4.1 BEE HONEY AND ITS BENEFITS FOR IMPROVING THE QUALITY OF HUMAN LIFE

The use of honey in medicine dates back to ancient times. Celsius was one of the great figures in medicine in the first century of the Christian era, and already stated that honey had a binding action on wounds (COSTA et al., 2006).

Honey is a very rich food with a high energy value, but its value is not only added to its high energy action, but especially to the enzymes, vitamins and the presence of important chemical elements for the proper functioning of the body, the trace elements. Honey contains most of the mineral elements essential for the human body, especially selenium, manganese, zinc, chromium and aluminium (SILVA, 2006).

Consumed worldwide and extremely important for the health of the human organism, when pure, it can have various properties: antimicrobial, healing, soothing, tissue regenerating, stimulating, among others (BIZZARIA, 2003).

According to Silva (2006), bee honey is a remarkably complex natural liquid containing around 180 substances. The composition of honey is largely made up of sugars, 38% fructose and 31% glucose. A large number of secondary compounds in honey are known to have antioxidant properties. These include phenolic acids, flavonoids, certain enzymes (glucose oxidase, catalase and peroxidase), ascorbic acid, hydroxymethylfurfuraldehyde, carotenoids, organic acids, Maillard reaction products, amino acids and proteins.

As it is made up of simple sugars such as glucose and fructose, its passage from the digestive tract into the bloodstream and from there into the cells where it is metabolized, does not require many transformations by juices, enzymes, etc., and its entry into cellular metabolism is relatively rapid (MENEZES, 2003).

According to Silva (2006), the properties of bee honey have led to it being incorporated into various food products, as well as being offered to and consumed by different population groups, given its therapeutic importance and its natural sweetening properties.

According to Costa et al. (2006), because the medium is a supersaturated solution of sugars, it has a low water activity and therefore does not offer favorable conditions for the growth of bacteria. In addition, the natural acidification of the medium can inhibit the development of many pathogens. The enzyme glucose oxidase, excreted by bees, is responsible for converting glucose, in the presence of water and oxygen, into gluconic acid and hydrogen peroxide, both of which are considered strong antioxidants that attack the envelope of microorganisms, preserving and maintaining the sterility of honey during ripening.

2.4.2 - HONEY AS AN IMPORTANT AGENT FOR SUSTAINABLE DEVELOPMENT.

Honey is considered to be the easiest bee product to exploit, the best known and the one with the greatest marketing possibilities. As well as being a foodstuff, it is also used in the pharmaceutical and cosmetics industries for its well-known therapeutic properties, so the honey agribusiness can be considered a way of achieving sustainable development in a region (FREITAS, 2004).

Beekeeping is a conservative activity and one of the few in the agricultural sector that fulfills all the requirements of the sustainability tripod: the economic because it generates income for farmers; the social because it uses family labor in the countryside, reducing the rural exodus; and the ecological because no deforestation is done to raise bees (GUIMARAES, 1986).

According to Jollivet (1994), beekeeping complements and benefits the other activities on the farm, as well as preventing fires and increasing the production of commercial crops through pollination.

Beekeeping is currently considered to be one of the greatest options, and can be considered the one that best remunerates the producer even in years of adverse weather conditions (SEBRAE, 2005).

2.5 The importance of nutritional education in developing healthy eating habits.

Nutritional education is conceptualized as an educational process in which, through the union of knowledge and experience of the educator and the student, the aim is to make people autonomous and confident in making their food choices in a way that guarantees a healthy and pleasurable diet, thus helping to meet their physiological, psychological and social needs (CAMOSSA et al., 2005).

Aiming to form or change healthy eating habits, this implies a huge change that is linked to the individual's daily practices and attitudes (BERNART; ZANARDO, 2011).

The activities carried out with pre-school children should be such that it is possible to understand them, taking into account their cognitive, motor, affective and other abilities, keeping the doctor/patient relationship healthy and respecting the individual characteristics of each group (SALVI, 2009)].

The most favorable environment for the development of nutritional education programs and actions is certainly the school. Because its structure is very close to the families of its students, it can also involve them, thus reaching a greater number of people involved in the patient's/student's social life. What's more, the cost-benefit ratio of school interventions is usually very good. (SALVI, 2009).

Educational games are essential for promoting good eating habits. Children learn more easily through play, especially when the subject of healthy eating includes increasing the consumption of fruit and vegetables and reducing the consumption of sweets, fried foods, etc. Schoolchildren come into greater contact with the most rejected foods (vegetables, fruit and legumes) and learn about the importance of consuming each food group (BERNART; ZANARDO, 2011).

3. Materials and methods

1.1 Type of Study:

The research was descriptive in nature, since it was based on the premise that problems can be solved and practices improved through the description and analysis of objective and direct observations. It was also exploratory, as its purpose is to show the possibility of a probable future event. According to Gil (2002), this type of research has as its main objective the improvement of ideas or the discovery of intuitions.

This study also took a quantitative and qualitative approach, as it used numerical measures to test the hypotheses and also examined more in-depth and subjective aspects of the subject under study, employing interpretative procedures with verbal representation of the data collected through open-ended questions of a qualitative nature (GUNTHER, 2006).

It was also participatory research, which according to Haguette (1999) has the following principles: the logical and political possibility of popular subjects and groups being the direct or associated producers of their own knowledge, which although popular is still scientific; the power to determine the use and political destination of the knowledge produced by the research, with or without the participation of popular subjects in its stages; and the place and forms of participation of erudite scientific knowledge and its professional agent of knowledge, in the 'work with the people' that generates the need for research, and in the research itself that generates the need for their participation.

According to Barbier (1996): participatory research is the definition of an intervention strategy based on building more democratic relationships between the actors.

1.2 Location and characterization of the study area

The study was carried out in an elementary school in the urban area of the municipality of Sao Domingos-PB, which according to the Geological Survey of Brazil - CPRM (2005) is located in the western region of the state of Paraiba, occupying an area of 227.2km2. The municipality was selected for the work because bee honey had never been offered in school meals and because it is an area suitable for the development of beekeeping.

The municipality was created by law n^0 5.902 of April 29, 1994 and installed on January 1,

1997. Agriculture is the community's main economic activity. There are five elementary schools in the municipality, one in the urban area and four in the rural area. The vegetation is basically Caatinga with stretches of deciduous forest. The climate is tropical semi-arid, with summer rainfall.

Figure 01: Location of the Municipality of Sao Domingos in Paraiba.

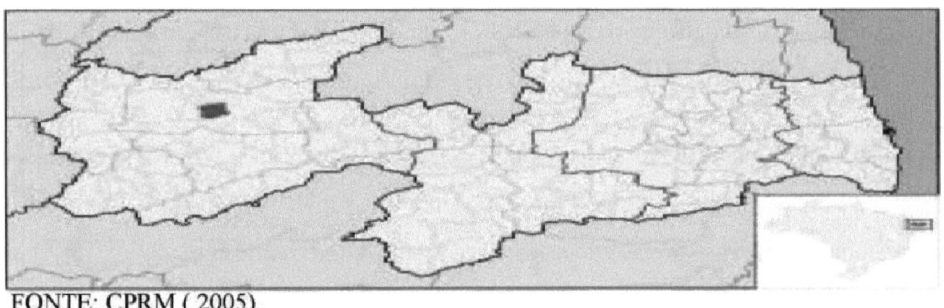

FONTE: CPRM (2005).

1.3 Population and sample

The research population consisted of the 180 students enrolled at the Maria Marques de Assis I Municipal Elementary School in the municipality of Sao Domingos, PB, the principal, 9 teachers and 2 cooks at the school, as well as the 10 rural family farmers in the municipality who supplied and/or supply food for school meals, the Constitutional Mayor and the Municipal Education Secretary of the state administrative district under study.

The sampling plan for the students was established in accordance with that proposed by Barbetta (2001) with a simple random sample selected from among them, with the number of students calculated with a tolerable sampling error of 4%. Thus, the number of students was equal to 60 children aged between 9 and 13, which represents 33.33% of the total number of students enrolled at the selected school.

Of the 60 students selected, 30 were studying in the morning shift and were chosen by lottery to receive nutritional education activities about honey. They were called GROUP 1 and the other 30 students were in the afternoon shift and only answered the questionnaires without receiving any clarification on the subject in question. They were called GROUP 2.

The sample was also made up of the population of teachers and cooks and the school principal, the population of rural family farmers in the municipality who supplied and/or supply food for school meals, plus the Mayor and the Municipal Secretary of Education.

1.4 Data collection procedures

1.4.1 QUESTIONNAIRE APPLIED TO STUDENTS BEFORE HONEY WAS ADDED TO THE MENU AT MARIA MARQUES SCHOOL

In September 2013, a semi-structured questionnaire (Appendix I) was given to a sample of students (GROUP 1 and GROUP 2) from the school under study, with simple and objective questions, in order to obtain a profile of the students in relation to their contact with honey outside of school, as well as to find out their intention to accept bee honey in their school meals.

1.4.2 DEVELOPMENT OF NUTRITIONAL EDUCATION ACTIVITIES WITH GROUP 1 STUDENTS.

Group 1 was made up of 30 students from the morning shift at the Maria Marques de Assis school who, throughout September 2013, received various educational activities related to honey, such as lectures held in the classroom and also in the school's audiovisual resources room, with the following themes: The life of bees and the formation of honey; The nutritional power of bee honey; The health benefits of honey.

The group also developed question and answer games, guessing games and word searches about honey in the classroom, as well as a cultural competition of drawings related to honey and its production and also about its benefits. At the end of the activities, the students were invited to write about what they had learned about honey.

Figure 02: Drawings that won prizes in the cultural drawing competition about the mei

Figure 03: Winning essays in the mei cultural essay contest

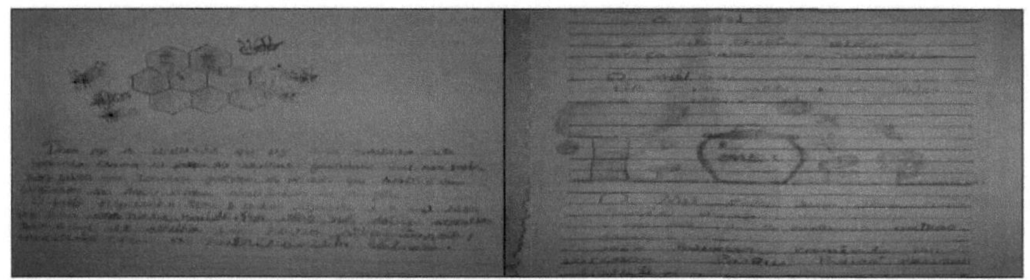

1.4.3 - IMPLEMENTING BEE HONEY IN SACHETS ON THE LUNCH MENU AT THE MARIA MARQUES DE ASSIS SCHOOL.

The municipality's nutritionist planned a menu to serve the half-apple in sachets as a dessert for the hot dog every Tuesday in October 2013. During that month, the planned menu was implemented and each student at the Maria Marques de Assis school received a sachet with 10g of half-apple as a dessert for the hot dog, following the proportions established by Pinheiro et. ai, (2005).

1.4.4 - APPLICATION OF THE QUESTIONNAIRE AFTER HONEY WAS OFFERED AS A SNACK AT THE MARIA MARQUES DE ASSIS SCHOOL.

In the questionnaire applied in the third week of October 2013, after the honey was offered as school food (Annex II), 2 simple closed questions and 2 questions using the five-point easy Hedonic scale established by the FNDE for testing acceptability were asked in order to detect the students' tolerance of bee honey as school food.

Figure 04: Five-point facial hedonic scale

I disliked it I didn't like it Indifferent I liked it I loved it

Source: FNDE, (2013).

The hedonic scale method is based on the conviction that direct responses, given on the basis of sensations, are more valid for predicting actual behavior towards food than responses that are more dependent on reasoning. Both the scale and the instructions are designed for use with individuals with entirely no experience in food testing (GARRUTI, 2003).

1.4.5 - SURVEY OF QUESTIONS WITH TEACHERS AND MEALERS AFTER HONEY HONEY WAS GIVEN IN THE MEAL OF THE MARIA MARIA MARQUES DE ASSIS SCHOOL.

At the end of October 2013, the nine teachers at the Maria Marques school and the two cooks at the same institution, one with 15 years of service at the school and the other with two years at the same place, answered a semi-structured questionnaire with closed questions, specific to each professional category (Appendices III and IV) in order to obtain the observations made by these professionals about the behavior of the students after receiving honey in their lunch, as well as to identify the importance given by these professional categories to the food under analysis.

1.4.6 - MEETING AND QUESTIONNAIRE WITH RURAL FAMILY FARMERS IN SAO DOMINGOS-PB.

On October 23rd, 2013, a meeting was held with rural family farmers from Sao Domingos-Pb, where the experience of introducing bee honey into the school meals of an educational establishment in the municipality was presented. It was quickly pointed out that if honey is definitively introduced into the school menu, the product will have a good demand and guaranteed sales, requiring trained local beekeepers to provide this quality food.

A semi-structured questionnaire with closed questions (Appendix 5) was administered to 10 local family farmers who sell or have sold food to the national school feeding program in Sao Domingos-PB.

1.4.7 - INTERVIEW WITH THE PRINCIPAL OF THE MARIA MARQUES SCHOOL, THE MUNICIPAL SECRETARY OF EDUCATION AND THE MAYOR OF THE MUNICIPALITY OF SAO DOMINGOS-PB.

The interviews were carried out in October 2013, after the students' acceptance of the honey had been confirmed. A questionnaire with open-ended questions was used, with one professional being interviewed at a time on different days (October 23, 24 and 25) and in different locations, and the responses of each were recorded for later analysis.

1.5 - Data analysis procedures

The information was analyzed using different methods. Descriptive statistics were used for the quantitative data, using the IBM SPSS Statistics software, version 21 of 2012.

The qualitative data was observed following the steps of the content analysis technique described by Bardin, which includes: first, reading each of the interviews; second, reading with the aim of breaking down the recording units in order to classify and group the raw data from the subjects' speeches into a representation of the content; and categorization by recording group with meaning and common elements (BARDIN,2OO9).

4. Results and discussion

Taking a closer look at the information found through this participatory research, we obtained :

4.1 From interviews with students in groups 1 and 2 before the introduction of honey in school meals:

Figure 05: Demonstration of honey consumption by students in Group 1 compared to those in Group 2.

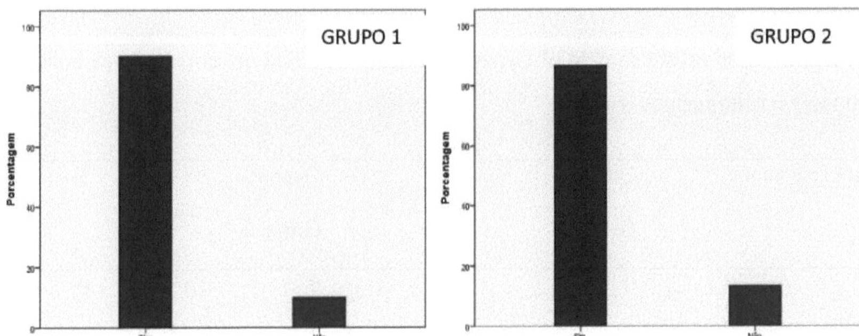

When asked if they had ever consumed bee honey in their lives, the two groups of students gave a similar percentage of answers: in Group 1, 90% of the interviewees had tried the food, while only 10% of them had never eaten bee honey; in Group 2, 86.7% of the students had consumed bee honey, while 13.3% had never tasted it.

The above results show that bee honey is a common food and confirm what Silva (2004) says, that honey is considered the easiest bee product to exploit and is also the best known.

Table 01: Distribution of the sample of Group 2 students who tasted bee honey in terms of whether they liked the taste of this food.

They liked the taste of honey	Quantity Group 1	%	Quantity Group 2	%
Yes	27	100	26	100
No	0			

Total	27	100%	26	100%

Table 02: Distribution of the sample of Group 2 students who have tasted bee honey in relation to their desire to include this food in their lunch.

They want honey in school meals	Quantity Group 1	%	Quantity Group 2	%
Yes	27	100	26	100
No	0	0	0	0
Total	27	100%	26	100%

Table 03: Distribution of the sample of Group 2 students who had never tasted bee honey in relation to their willingness to try it.

They want to try honey	Quantity Group 1	%	Quantity Group 2	%
Yes	0	0	0	0
No	3	100	4	100
Total	3	100	4	100%

Looking at the tables above, it can be seen in table Ol that 100% of the students in the two groups who have already eaten bee honey liked its taste and in table 02 that the same percentage of students in both groups are eager to introduce bee honey into school meals, demonstrating that it is a food that is easily accepted, However, table 03 shows that the students in both groups who had never consumed bee honey also had no intention of trying it, confirming the claim that children are naturally predisposed to reject a new food (neophobia) in their eating habits (VITOLO, 2008).

Table 04 Distribution of the sample of students (le2 groups) who have never received bee honey as part of their school meals during their school years.

They received the honey in the lunchbox	quantity	%
Yes	0	0

No	60	100
Total	60	100%

Table 04 shows that there was no bee honey on the school menu in Sao Domingos-PB before this work was carried out, because although the research was carried out in one school out of five, the menus are standardized and offered equally in all schools.

4.2 From the questionnaire applied to the sample of students (Group 1 and 2) after bee honey in sachets was offered in school meals.

Figure 06: Students' opinions (groups l and 2) on the use of bee honey in snacks

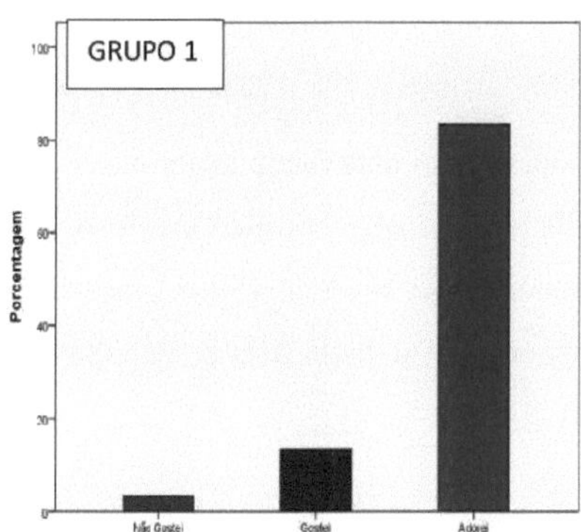

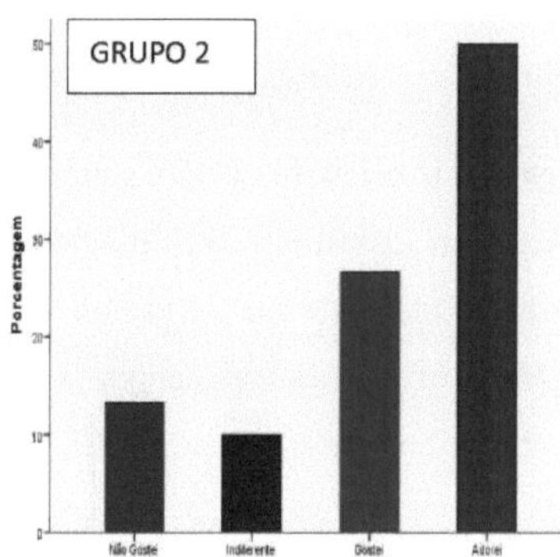

It can be seen from the figure above that the graph representing the acceptability of GROUP l, the group that received educational activities on bee honey, showed a better satisfaction profile when receiving honey in school meals, with 83.3% of the sample saying that they loved the presence of honey in their lunch, 13,3% said they liked it and only 3.3% didn't like the honey, representing an acceptability rate for this food of 96.6%, indicating that for this group this food can be introduced into the school meals under study, as the minimum acceptance rate for foods set by the FNDE is at least 85%.

According to Magalhaes (2005), the acceptance of food by students is the main factor that determines the quality of the service provided by schools when it comes to supplying school meals. According to the PNAE rules, educational institutions must apply an acceptability test whenever there is a food on the menu that is atypical of the school's eating habits, and it can be offered permanently as long as the acceptability index is not less than 85% (eighty-five percent).

In addition to building important knowledge about food and nutrition, educational actions encourage and improve the frequency of healthy eating practices. Educational dynamics are essential in promoting good eating habits (BERNART; ZANARDO, 2011).

In figure 06, the graph representing the acceptability of bee honey in school meals by GROUP 2, the group that did not receive any type of educational activity, reflects that 50% of the students in the group loved consuming the honey in school meals and 26.7% liked it, but 10% of the sample was indifferent to the offer of honey and 13.3% did not like it,3% did not like the food on the menu, which shows a percentage of acceptability of 76.7%, that is, an acceptability lower than the minimum required by the FNDE to introduce a food on the school lunch menu, so for group 2, it would not be possible to offer honey in the school under study.

Although this is below the FNDE's proposal, it is important to note that a major obstacle to increasing honey consumption is the fact that Brazilians, in general, consider this food to be just a natural medicine that is useful for the respiratory tract and that without the proper development of educational actions to show that honey is a food rich in nutrients, its acceptance will not be so tolerant (EBELING, 2002).

Figure 07: Opinion of the students in Groups le2 on the way the honey was offered (in sachets)

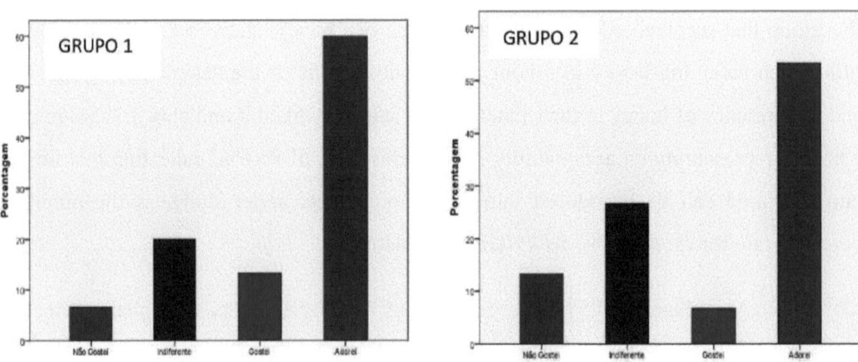

The students' opinions on how to receive the honey (in sachets) were similar when comparing the two groups. In GROUP 1 60% of the sample loved the honey in a sachet and 13.3% liked it that way, 20% were indifferent to the sachet and 6.7% didn't like receiving the honey in a sachet. In GROUP 2 53.3% of the students loved the honey packaged in a sachet and 6.7% liked it, while 26.7% were indifferent and 13.3% didn't like it.

According to Scherer (2006), an important action taken to introduce bee honey into school meals is to offer it in sachets, with daily consumption estimated at 10g/student/day.

However, a considerable percentage did not like the sachet, especially in group 2, which did not receive educational activities. It should be noted that the percentage of students who did not like honey in sachets was the same as the percentage of the group who had never tried honey and also had no desire to try it and without awareness of the importance of food, even if they did try it, they still did not like honey, not even the way it was presented.

Table 05 Sample of students (groups le 2) who would like mei to be offered more often on the school lunch menu.

They want honey in their lunchboxes more often	Quantity Group 1	%	Quantity Group 2	%
Yes	30	100	26	86,7
No	0	0	4	13,3
Total	30	100	30	100

Examining the data in table 05 shows that the vast majority of students want honey to be offered more often on the school menu, especially those in group l, where 100% of the sample said they wanted honey to be included more often in school meals. 3.3% of the students in this same group said in another question that they didn't like honey in school meals, but they were aware of the importance of the food when they answered that even if they didn't like it, they wanted honey to be included more often on the school menu.

In group 2, 13.3% of the sample did not want honey to be served more often in school meals. This is the same percentage of pupils in this group who also said they did not like honey before and after its implementation in school meals, which means that because they have not had any activities to make them aware of the importance of bee honey in school meals for their health and also for local economic development, they are not interested in introducing this foodstuff into school meals.

Table 06: Sample of students (groups le 2) who consider honey to be an important food for human health

They want honey in their lunchboxes more often	Quantity Group 1	%	Quantity Group 2	%
Yes	30	100	20	66,7
No	0	0	10	33,3
Total	30	100	30	100

Looking at the data in the table above, 100% of the students in group 1 consider honey to be an important food for human health. This also highlights the importance of knowledge about the food, because even though 3.3% of them don't like honey, they recognize that it is important for health. In group 2, 33.3% of the students believe that honey has no health value, a higher percentage than the 13.3% of students in this same group who don't like honey, which means that even though some of the students like bee honey, they don't consider it important.

There is a real need, among other measures, for food education programs that promote honey as a healthy and interesting food, in order to improve its acceptability and awareness of its importance (MAGALHAES, 2005).

4.3 The questions asked of the lunch ladies at the Maria Marques de Assis School after honey was added to their school meals.

When we asked these professionals, we realized that they had never served bee honey in school meals in Sao Domingos-PB, even though one of them has worked as a lunch lady at the school under study for fifteen years, i.e. she has been a lunch lady since the school was founded. We also noticed that the two lunch ladies consider honey to be an important food and that it should be included in school meals in the municipality. According to them, when they carefully examined the students' behavior when they received the honey in sachets, they considered that this food was well accepted by them.

From this information, it is clear that although honey has never been introduced into school meals in Sao Domingos-PB, it has two fundamental allies in its inclusion on the menu of school meals in the municipality, as they consider this food to be important for health,

According to Gonçalves (2006), the fact that the cooks like the food that is being included in school meals makes it easier for them to accept it, since they tend to pass on their tastes and habits in a positive and/or negative way when they offer it.

When students receive the food, most of them don't think about the nutrients it contains, and with lunch ladies who consider the food offered to be important and/or tasty, the way it is prepared and distributed can be influenced, since during these processes it is essential that lunch ladies express love and affection for what they do (GONÇALVES, 2006).

4.4 The questions asked of teachers at the Maria Marques de Assis School after honey was introduced into their school meals.

. The teachers unanimously said that they had observed a good acceptance of the sachet of bee meat by the pupils and that they also considered this food to be important for the human organism and that offering it in school meals could promote the health and improve the development of their pupils, expressing the desire for bee meat to be a permanent item in the school meals of Sao Domingos-PB.

Teachers' good view of honey is of great value when it comes to implementing it in school meals, because according to Gonçalves (2006), everyone at school should have a common goal: the education of students, which includes food education.

The formation of food preferences in children is part of a learning process in which they tend to imitate what others do, whether they are family members or other people of reference, including teachers who play a fundamental role in nutritional learning, as their knowledge, attitudes and practices have a decisive influence on students (JOMORI et. al, 2008).

4.5 From the series of questions applied to rural family farmers in Sâo Domingos, PB.

After attending a meeting on beekeeping, all ten rural family farmers in the municipality of Sâo Domingos who provide and/or have provided food for school meals said that they intended to receive specific training to become local beekeepers. They also said that if they were currently honey producers, they would be able to earn a higher family income and also said that a beekeeper in the municipality of Sâo Domingos would have a certain market in the region to sell their products.

The answers to the questionnaires show that farmers are very interested in cultivating bees and producing honey, since the PNAE would provide them with a good quantity of honey with guaranteed sales. However, in order for this to happen, they must receive training in the area, have equipment and tools available to practice beekeeping and Sao Domingos must have a Municipal Inspection System to certify the quality of the honey produced, centrifuged and packaged.

The interest of these farmers is a very important factor for the economic development of the municipality and for improving the quality of life of the people of Saodomingu.

According to Magalhaes (2005), the inclusion of honey in school meals should become an opportunity for the beekeeping agribusiness, guaranteeing a fair price for the producer and more accessibility for the consumer, creating a culture of consumption in school and family meals.

Exploitation of the domestic market must take place within socio-productive configurations that provide a competitive advantage for the honey production chain, such as the Local Agri-Food System (SIAL), the space of which, in addition to its geographical content, is constructed by collective actions, marked by cultural issues and institutionally regulated. What takes place is an interaction between the territory and the production chain (production-distribution-consumption) of a given foodstuff. This space for cooperation creates opportunities and new ways for producers to operate in the competitive environment, mainly by reducing transaction costs and the high specificity of the assets present in the territory (LINS, 2004).

4.6 Interviews with the school principal, the secretary of education and the constitutional mayor of the municipality of Sâo Domingos-PB

If we look at the results of the study up to the day of the interview with each of the professionals mentioned above, we can see that they generally felt that the research had made a great contribution to the Maria Marques School and to the municipality of Sao Domingos as a whole, by showing that with nutritional education it is possible to achieve good acceptability of honey among schoolchildren. In their reports, the interviewees always highlighted honey as a natural food that has many riches to offer the students.

When asked about the possibility of permanently adding bee honey to the school menu, the school principal said she was in agreement, as she was able to closely monitor the work carried out in the study and liked the students' acceptance of this food. The Department of Education and the Mayor of the municipality said that, given the results obtained at the Maria Marques School, they approved the permanent inclusion of bee honey on the menu, not only at the school under study, but also at all the municipal primary schools.

When asked if they approve and support the training of rural family farmers to become local beekeepers, they were very interested, saying that they approve and support this procedure.

The principal of the school studied is also the president of the CMDRS (Municipal Council for Sustainable Rural Development) and so, as a form of support, she said she would mobilize farmers from all the rural community associations whenever necessary, as well as guaranteeing that from that moment on she would highlight at every council meeting the importance of the sustainable development generated by the production and marketing of honey in the region.

As well as approving the training of beekeepers, the Mayor assured that she would support them as much as possible, either through agreements with institutions that provide training or by purchasing specific materials, He also promised to put the project for the creation of the Municipal Inspection System (SIM) in Sao Domingos-Pb to a vote in the City Council and then to make it effective so that rural family farmers in Sao Domingos can sell animal products, such as honey, to the PNAE and thus increase their income.

The Municipal Secretary of Education, as head of the PNAE executing entity, says that the government requires that at least 30% of the financial resources passed on by the FNDE be used to buy food from family farms, but that she would be willing to allocate up to 100% of the resources passed on by the FNDE for school meals to the purchase of bee honey from future local beekeepers.

After the end of this study, the Municipal Inspection System (Annex VII) was approved and created in the municipality, as promised by the mayor during the interview. This system is currently being installed with the support of the municipal health department, which already has a specific room for its operation and a veterinarian, a professional who is currently in charge of organizing the records and carrying out the tasks required for the system to come into force in the municipality of Sao Domingos-PB.

According to Magalhaes et. al, (2005), the current moment in beekeeping has called for a change in behavior, where various sectors need to intensify and support debate, reflection and action on a competitive strategy that unites the efforts of the main players (producers, consumers, development entities) to strengthen the Brazilian beekeeping sector, especially if efforts are concentrated on solving some of the bottlenecks in the honey production chain, such as the absorption of its production.

5. FINAL CONSIDERATIONS

On the basis of the results obtained, examined and discussed in this study, it can be seen that:

The students at the school under study had never actually received honey from bees in their school meals and the two groups of students studied (Group 1 with educational activities on honey, Group 2 without educational activities on honey) were similar in terms of their contact with honey outside of school and their intention to receive it in their school meals before the application of nutritional education on honey and its introduction;

The students who received educational activities about bee honey and its benefits showed better acceptability of the food than those who did not. The group that received nutritional guidance about honey showed an acceptability index of 96.6%, well above the minimum required by the FNDE, which is 85% for a food to be included in school meals, while the group of students without educational activities had an acceptability index of 76.7%, a good percentage, but lower than the minimum required for schoolchildren, showing that bee honey is a food that is easily accepted, but to be introduced as a food in a school it needs prior educational work;

Of the group of students who took part in the nutritional education on honey, some (10% of them) didn't like honey even though they were made aware of it, but because of their knowledge of the benefits of this food, 100% of them reported that they wanted to know that bee honey is very important for health and that it needs to be included in school meals;

The form in which the honey was offered (in sachets) showed good acceptability in both groups, although it was slightly higher in the group that received nutritional education;

Food education has contributed to the formation of healthy eating practices, helping to include natural, healthy and unusual foodstuffs in the school menu.

It is of great importance that the entire school community (principal, teachers and cooks) have a good view of the food offered in school meals;

The rural family farmers of Sao Domingos-Pb have realized the importance of beekeeping for the region and have woken up to developing this practice.

The research has brought about real and important changes for the municipality of Sao Domingos-Pb, as the study has shown real attitudes that could lead to sustainable local development, as it has encouraged the Constitutional Mayor with the support of the education and health secretariats to set up a Municipal Food Inspection System, as well as entering into a partnership with SEBRAE (Brazilian Service to Support Micro and Small Enterprises) to teach a course on good food production and manufacturing practices.

Finally, it is essential that other studies are carried out along the same lines, with the aim of improving the quality of school meals and promoting the economic, social and health development of people in the region.

6. References

ACCIOLY et. al. **Nutrition in Pediatrics and Obstetrics. Guanabara Koogan**, 2ª Ed.Rio de Janeiro, 2012.

BARBETTA, P. A. **Estatistica aplicada às Ciências Sociais**. 4 ed.Florianôpolis: UFSC, 2001. 338p.

BARBIER, R. **La recherche action**. Ed. Anthropos/Economica - Paris, 1996. 112p

BARDIN, L. **Content Analysis**. 70 ed. Lisbon: [Sn], 2010-2009.

BERNART, A.; ZANARDO, V.P.S. Nutritional education for children in public schools in Erechim/RS. Revista Eletronica de Extensao da URI. v.7, n.13, p.71-79, 2011.

BIZZARIA, D. K.; FILGUEIRAS, C. T. Microbiological analysis of bee honey consumed in the municipality of Campo Grande-MS. **Hig. Alim.**, v. 17, p. 104-105, 2003.

BRAZIL. Ministry of Health. **Food guide for the Brazilian population: promoting healthy eating**. Brasilia: Ministry of Health, 2005. 236p.

BRAZIL. **Normative Instruction n.⁰ 11**, of October 20, 2000. Official Gazette, October 23, 2000. Section 1, p.16-17. Technical regulation on the identity and quality of honey.

CAMOSSA et. al.Nutrition EducationAn area in development. **Alimentos e Nutriçâo** Araraquara. v.16, n.4, p. 349-354, Oct./Dec. 2005.

CONSEA - NATIONAL COUNCIL FOR FOOD AND NUTRITION SECURITY. **Principios e diretrizes de uma política de segurança alimentar e nutricional: textos de referência da II Conferência de Segurança Alimentar e Nutricional.** Brasilia, July 2004. Available at: www.presidencia.gov.br/consea. Accessed on August 10, 2013

COSTA.J.M.C et al. Composition and therapeutic properties of bee honey. **Rev. Alim. e Nutr.** Araraquara. v.17,n.1, p.113-120.jan/mar. 2006.

CPRM. **Diagnosis of the Municipality of Sao Domingos de Pombal**. Project to Register Groundwater Supply Sources in the State of Paraiba. Recife, 2005. Available at: http://www.cprm.gov.br/rehi/atlas/paraiba/relatorios/SAOD174.pdf. Accessed on August

19, 2013.

EBELING, E. Beekeeping. In: Congresso brasileiro de apicultura, 14., 2002 Campo Grande, MS. **Proceedings**. Campo Grande: CBA: UFMS: FAAMS, 2002. p.166.

FNDE. **School meals**. Available at: http://www.fnde.gov.br/index.php/programas-alimentacao-escolar. Accessed on January 3, 2013

FRIEDMANN, Harriet. A sustainable world food economy. In: BELIK, L; MALUF, R.S. **Abastecimento e Segurança Alimentar.** Campinas: UNICAMP, 2000. p. 121.

FREITAS, D. G. F.; KHAN, A. S.; SILVA, L. M. R. Technological level and profitability of honey bee production (Apis mellifera) in Ceara. **Rev. Econ. Sociol. Rural**, v. 42, n. 1, p. 171-188, 2004.

GARRUTI, D. Sensory profile and acceptance of "requeijao cremoso" cheese. **Rev. Cien. Tecnol. de Alimentos.** Vol 23, n. 3, pp 434-440. Campinas. Sep-Dec 2003.

GUIMARAES, N. P. **Beekeeping, the science of long life**. Ed.Itatiaia Ltda. Belo Horizonte. 1986.

GIL, A. C. **Como elaborar projetos de pesquisa**. 4 ed. Sao Paulo: Editora Atlas S.A., 2002. ch. 4, p.41-56

GONÇALVES, L. C. **Politicas de Alimentaçao Escolar. Brasilia**. 1 ed Centro de Educaçao a Distància - CEAD, Universidade de Brasilia, 2006.

GÜNTHER, H. Qualitative Research Versus Quantitative Research: Is This the Question? **Psicologia: Teoria e Pesquisa,** Sciello, vol. 22 n^{0} 2, p.201-210 May/Aug. 2006.

HAGUETTE, T. M. F. **Metodologias Qualitativas na Sociologia**. 6 ed. Petrópolis: Vozes, 1999. 224p.

JAMORI et. al. Dietary choice determinants. **Revista de Nutriçâo**. Vol. 21, n^{0} 2. April/2.008

JOLLIVET, M. Agricultura e meio ambiente: reflexoes sociológicas. In: **Estudos Economicos**: Sao Paulo, USP. Vol. 240, no especial, p.183-198, 1994.

KAC;VELASQUEZ-MELÉNDEZ,G.**The nutritional transition and the epidemiology of obesity in Latin America.** Cad. Saùde Pùblica, 19 (Suppl. 1):2003

LINS, H. Territory, culture and innovation: the perspective of localized agri-food systems. In: Encontro Nacional de Economia Politica, 9, 2004, Uberlândia. **Proceedings...** Uberlândia: SEP, 2004.

MAGALHÂES et al.Feasibility of introducing honey into school meals: opportunity and challenge for the beekeeping agribusiness. **Revista de economia e agronegócio.** Vol.7 n. 1. Florianópolis. UFSC, 2005.

MENEZES, P. Bee honey, medicine or food? **Rev.Mensagem Doce**, v. 73, 2003.

Available at: http://apacame.org.br/mensagemdoce/73/comentario.htm. Accessed on: August 19, 2013.

PERONDI, M A.; RIBEIRO, E. M. As estratégias de reproduçao de sitiantes no oeste de Minas Gerais e de colonos no sudoeste do Parana. **Rural and Agroindustrial Organizations.** V.2, n.2, July-December, 2000.

National SEBRAE. **The experience of the apis network**. Results-oriented management, 2005.

SALVI, C.; CENI, G.C.. Nutritional education for preschoolers at the Mother Alix nursery school association. Experiences: Revista Eletronica de Extensao da URI. v.5, n.8, p.71-76, October/2009.

SILVA.R. A et al. Composition and therapeutic properties of bee honey. Ver. **Alim.**

Nutr. Araraquara, v.17, n.1, p.113-120,jan./mar. 2006.

SCHUCH, H. J. **The importance of opting for family farming**. Available at: http://gipaf.cnptia.embrapa.br/itens/publ/fetagrs/fetagrs99.doc Accessed on: November 20, 2013.

SOUTO,J.W. **Contribution to project pedagogy**, 2011. Available at: <http.www.google.com.br>.Accessed on August 20, 2013

SCHERER, W. **Mei in school meals**. Porto Alegre, UFRGS, June 9, 2006. Lecture given at the Seminar on Agribusiness in Beekeeping of the PPG-Agronegocios of Cepan/UFRGS.

PINHEIRO, et. al. **Table for evaluating food consumption in home measures.** Rio de Janeiro: 2005.

PHILIPPI, S. T. **Food composition table: support for nutritional decisions**. Brasilia: UnB, 2001. 133p

WANDERLEY. M.N.B. A Valorizaçao da agricultura familiar e a reinvindicação da ruralidade no Brasil**. Rev. Desenvolvimento e Meio Ambiente.**UFPR. v.2,n.2, 2000.

VITOLO, M.R. **Nutrition from pregnancy to ageing**. Ed Rubio. Rio de Janeiro, 2008.

7. ANNEXES

Annex I - Questionnaire applied to students before the implementation of honey in 2013

Student:Class:Year.

1- Do you consume or have you consumed bee honey as food?

()YES ( )NO

2- If Not, Do You Feel Like Trying It?

()YES ( )NO

3- If so, did you like the taste of the honey?

( ) I liked it ( ) I didn't like it

4- Do you consume or have you consumed bee honey in school meals?

()YES ( )NO

5- If yes, would you like bee honey to be offered more often?

( )YES( )NO

6- If No, would you like to have bee honey in your school lunch?

( )YES( )NO

Annex II - Acceptability test applied to students after the implementation of honey in 2013.

Student:Class:

1- What did you think of offering mei as a dessert in school meals?

12345

Disliked Disliked Indifferent

LikedLiked

2-0 what about the way it was offered (in a sachet)?

12345

Disliked Disliked Indifferent

LikedLiked

3 - Would you like honey to be included in your school lunch more often?

( ) Yes ( ) No

4 - Do you consider honey an important food for your health?

( ) Yes ( ) No

Annex III - Questionnaire applied to school cooks in Sao Domingos-PB after the introduction of bee honey in school meals.

Catering: ____

1- How many years have you worked as a lunch lady in the city of sao domingos?

2- Have you ever served honey in school meals in São Domingos before?

()Yes( )No

3- Do you think honey is an important food to include in school meals?

( ) yes ( ) no

4- When you serve the honey, how do you assess the students' acceptance when you observe the body expressions and comments made by the students during the delivery of the snack?

( ) Horrible ( ) Bad ( ) Fair ( ) Good ( ) Great

Annex IV - Questionnaire applied to teachers.

Professerà: ______________________________________

1 - Do you consider bee honey to be an important food for the human body?

( ) yes ( ) no

2 - Do you think that honey from bees can promote the development and health of your students?

( ) yes ( ) no ( ) can't say

3 - Would you like bee honey to be a permanent item in school meals?

( ) yes ( ) no ( ) whatever

4 - From your students' comments in class, have you noticed a good acceptance of the bee honey you serve them?

( ) yes ( ) no ( ) I didn't understand

Annex V - Questionnaire applied to rural family farmers

Farmer: ______________________________

1 - Would you like to receive specific training to become a honey producer?

( ) yes ( ) no

2-If you were a honey producer today, do you think your family income would be higher?

( ) yes ( ) no

3 - As a honey producer in Sao Domingos, do you think you'd have the right market for the

honey you produce?

( ) yes ( ) no

Annex VI - Questionnaire applied to the Constitutional Mayor of the Municipality of Sao Domingos, the Municipal Education Secretary and the School Director.

Person interviewed: ______________________

1- What do you think of the project to introduce bee honey to primary schools in your municipality?

2- Observing the acceptability of bee honey by the students, do you intend to permanently introduce this food into the lunchbox?

3- Do you approve and guarantee support for training local farmers to become beekeepers?

4- Once the farmers produce the honey, would you buy local produce to use in the school lunch?

5- Do you think that incorporating bee honey into school meals and buying it from local farmers can generate health, economic and social improvements for the local population? Why?

Appendix 01

- Law № 293/2013 of October 29, 2013

STATE OF PARAIBA

CITY HALL OF SAO DOMINGOS

MAYOR'S OFFICE

LEI № 293/2013, of October 29, 2013.

PROVIDES FOR THE CREATION OF THE MUNICIPAL INSPECTION SERVICE (SIM) AND SETS STANDARDS FOR THE INSPECTION SERVICE AND THE HEALTH INSPECTION PROCEDURES OF ESTABLISHMENTS THAT PRODUCE FOOD FOR HUMAN CONSUMPTION OF ANIMAL AND VEGETABLE ORGANIZATION IN THE MUNICIPALITY OF SAO DOMINGOS-PB AND PROVIDES OTHER PROVISIONS.

THE CONSTITUTIONAL MAYOR OF THE MUNICIPALITY OF SAO DOMINGOS

I hereby inform you that the City Council approves and I sanction the following Law:

Γ - The Municipal Inspection Service (SIM) is hereby established, subordinate to the Municipal Secretary of Agriculture, whose purpose is the industrial and sanitary inspection of products of animal origin, both edible and non-edible, whether or not they contain added vegetable products, prepared, processed, handled, received, packaged, deposited and in transit in the Municipality of Sao Domingos - Paraiba.

Sole Paragraph - This Law complies with Federal Law No.^{0} 9.712/1998 and Federal Decree No. 5.741/2006, which established the Unified Agricultural Health Care System (SUASA).

Art. 2 - The sanitary inspection of beverages and foodstuffs for human consumption of animal and plant origin refers to the systematic process of monitoring, evaluation and sanitary control, from the raw material to the preparation of the final product and will be the responsibility of the Municipal Inspection Service (SIM), a body subordinate to the Municipal Secretary of Agriculture of the municipality of Sào Domingos-Paraiba.

§ 1^{0} - The presence of the inspector in establishments is compulsory when animals are slaughtered, in the case of a slaughterhouse, for the pre- and post-mortem inspection of animals and carcasses.

§ Paragraph 2 - The permanent presence of inspectors at

CNPLOI .612.691 /0001 -47 Rua Projetada. s/n. CEP.58.853-000 TEL(OSS) 3432-1003

Printed by Books on Demand GmbH, Norderstedt / Germany